The

Chronicles

Of

Iris

The

Chronicles

Of

Iris

...futuristic predictions

Cosmic Predictions
Volume 1
Copyright 2020

To

those

who

believe

Cosmic Predictions

Volume #1

Table of Contents

Cosmic Predictions
Volume #1
Table of Contents

Cosmic Predictions

Volume #1

Table of Contents

The Chronicles of Iris

...futuristic Predictions

Cosmic Predictions

Volume 1

Disclaimer

Chronicled Futuristic Predictions are
intuitive opinions made by the Author.
This book is protected under the
First Amendment of United States
Constitution.
For Entertainment Purposes Only
ReadersViewer discretion advised

Anatomy of a Chronicled Prediction

Origin of the Prediction:

The Chronicles of Iris

...futuristic predictions

Chronicle number: (01012018-A)

(month, day, year, letter)

Date of Chronicle Prediction: January 1, 2018

Prediction Number; Prediction 1

Prediction: "Ocean temperatures continue to rise

The Chronicles of Iris

...futuristic predictions

Cosmic Predictions
Volume 1

The Chronicles of Iris

...futuristic predictions

Chronicle (11012018-A)

November 01, 2018

Prediction 1

"Spacecrafts fueled by Cosmic Rays"

The Chronicles of Iris

...futuristic predictions

Chronicle (11012018-B)

November 01, 2018

Prediction 2

"Mercury's interior has flowing liquid water"

The Chronicles of Iris

...futuristic predictions

Chronicle (11012018-C)

November 01, 2018

Prediction 3

"Mercury's interior is a candidate for the sustainment of Life"

The Chronicles of Iris

...futuristic predictions

Chronicle (11012018-D)

November 01, 2018

Prediction 4

"Mercury's planetary Orbiter malfunctions"

The Chronicles of Iris

...futuristic predictions

Chronicle (11012018-E)

November 01, 2018

Prediction 5

"A Rover lands whilw on a mission to Mars"

The Chronicles of Iris

...futuristic predictions

Chronicle (11022018-F)

November 02, 2018

Prediction 6

Anti-gravity interdimensional
Spacecraft built for traveling
into Black Holes, Worm holes
and White Holes"

The Chronicles of Iris

...futuristic predictions

Chronicle (11022018-G)

November 02, 2018

Prediction 7

"Colossal asteroid smashes into Mercury"

The Chronicles of Iris

...futuristic predictions

Chronicle (11022018-H)

November 02, 2018

Prediction 8

Specially designed advanced (Ai) Spacecraft is built for Venus"

The Chronicles of Iris

...futuristic predictions

Chronicle (11022018-I)

November 02, 2018

Prediction 9

A new mission to

Venus

discovers Oxygen"

The Chronicles of Iris

...futuristic predictions

Chronicle (11022018-J)

November 02, 2018

Prediction 10

"Record breaking
Sun flare interferes
with communication
thru out the Milky Way

The Chronicles of Iris

...futuristic predictions

Chronicle (11022018-K)

November 02, 2018

Prediction 11

"Bio-organisms

discovered on

Jupiter's moon, Europa"

The Chronicles of Iris

...futuristic predictions

Chronicle (11022018-L)

November 02, 2018

Prediction 12

"Life exists in the

subsurface

of Luna, Earth's moon"

The Chronicles of Iris

...futuristic predictions

Chronicle (11022018-M)

November 02, 2018

Prediction 13

"Enceladus' interior reveals a platform for sustainable Life"

The Chronicles of Iris

...futuristic predictions

Chronicle (11032018-N)

November 03, 2018

Prediction 14

"Interior of Jupiter's moon, Europa reveals sustainable elements necessary for Life"

The Chronicles of Iris

...futuristic predictions

Chronicle (11032018-O)

November 03, 2018

Prediction 15

"*Living organisms detected within Saturn's moon, Enceladus*"

The Chronicles of Iris

…futuristic predictions

Chronicle (11032018-P)

November 03, 2018

Prediction 16

"Orbiter travels to Saturn's moon Enceladus searching for the existence life"

The Chronicles of Iris

...futuristic predictions

Chronicle (11032018-Q)

November 03, 2018

Prediction 17

An (AI) Rover
lands on the
surface of Venus"

The Chronicles of Iris

...futuristic predictions

Chronicle (11042018-R)

November 04, 2018

Prediction 18

"Flyby mission to research Enceladus' atmosphere"

The Chronicles of Iris

...futuristic predictions

Chronicle (11042018-S)

November 04, 2018

Prediction 19

"Subterranean oceanic Volcanoes feed the geysers blasting on Enceladus, Saturn's moon"

The Chronicles of Iris

...futuristic predictions

Chronicle (11042018-T)

November 04, 2018

Prediction 20

"Space station built
on Saturn's moon
Enceladus"

The Chronicles of Iris

...futuristic predictions

Chronicle (11042018-U)

November 04, 2018

Prediction 21

"Habitable exoplanets discovered beyond Pluto"

The Chronicles of Iris

...futuristic predictions

Chronicle (11052018-V)

November 05, 2018

Prediction 22

"Active Caldera
uncovered
on Venus"

The Chronicles of Iris

...futuristic predictions

Chronicle (11052018-W)

November 05, 2018

Prediction 23

"Newly uncovered

active

Volcanoes on Venus"

The Chronicles of Iris

...futuristic predictions

Chronicle (11052018-X)

November 05, 2018

Prediction 24

"Water vapor traces detected in Venus' upper cloud level"

The Chronicles of Iris
...futuristic predictions

Chronicle (11052018-Y)

November 05, 2018

Prediction 25

"A balance between carbon dioxide and oxygen is reached in Venus' atmosphere"

The Chronicles of Iris

...futuristic predictions

Chronicle (11052018-Z)

November 05, 2018

Prediction 26

"Venus proves

to be

habitable"

The Chronicles of Iris

...futuristic predictions

Chronicle (11062018-A1)

November 06, 2018

Prediction 27

Prediction 27

"Mars displays
signs of more
pre-existing lifeforms'

The Chronicles of Iris

...futuristic predictions

Chronicle (11062018-B1)

November 06, 2018

Prediction 28

"The Earth's moon Luna is habitable"

The Chronicles of Iris

...futuristic predictions

Chronicle (11062018-C1)

November 06, 2018

Prediction 29

"Land Probe survives l
anding on
Venus and transmits data"

The Chronicles of Iris

...futuristic predictions

Chronicle (11062018-D1)

November 06, 2018

Prediction 30

"Mars has lifeforms dwelling on the interior"

The Chronicles of Iris

...futuristic predictions

Chronicle (11062018-E1)

November 06, 2018

Prediction 31

"Space station

built on the

Earth's moon, Luna"

The Chronicles of Iris

...futuristic predictions

Chronicle (11062018-F1)

November 06, 2018

Prediction 32

"Spacecrafts operating on a new energy discovered in Space"

The Chronicles of Iris

...futuristic predictions

Chronicle (11062018-G1)

November 06, 2018

Prediction 33

"Space station

built

on Mars"

The Chronicles of Iris

...futuristic predictions

Chronicle (11112018-H1)

November 11, 2018

Prediction 34

"Spacecraft explodes on the launching pad in Japan"

The Chronicles of Iris

...futuristic predictions

Chronicle (11112018-I1)

November 11, 2018

Prediction 35

"Spacecraft explodes
after launching
from Russia"

The Chronicles of Iris

…futuristic predictions

Chronicle (11122018-J1)

November 12, 2018

Prediction 36

"Cosmic Ray propulsion system for Interdimensional spacecrafts and Space stations"

The Chronicles of Iris

...futuristic predictions

Chronicle (11122018-K1)

November 12, 2018

Prediction 37

"Certain organs physical
are replicated by
isolating the Dna sequence"

The Chronicles of Iris

...futuristic predictions

Chronicle (11112018-L1)

November 11, 2018

Prediction 38

"Human children

created from two

individual strands of Dna code

(No sperm or eggs required)

The Chronicles of Iris

...futuristic predictions

Chronicle (11112018-M1)

November 11, 2018

Prediction 39

"A moving Light
tunnel to and
from the Earth
and to other planets"

The Chronicles of Iris

...futuristic predictions

Chronicle (11132018-N1)

November 13, 2018

Prediction 40

"Venus' possess

an internal

Water supply"

The Chronicles of Iris

…futuristic predictions

Chronicle (11132018-01)

November 13, 2018

Prediction 41

"Algorithmic mathematics applied to Dimensional space travel"

The Chronicles of Iris

…futuristic predictions

Chronicle (11132018-P1)

November 13, 2018

Prediction 42

"Persistent destructive pounding Space weather smashes into Earth's moon Luna"

The Chronicles of Iris

...futuristic predictions

Chronicle (11132018-Q1)

November 13, 2018

Prediction 43

"Obliterating interplanetary weather pounds Mercury"

The Chronicles of Iris

...futuristic predictions

Chronicle (11132018-R1)

November 13, 2018

Prediction 44

"A gargantuan sudden explosion in Saturn's ring"

The Chronicles of Iris

...futuristic predictions

Chronicle (11132018-S1)

November 13, 2018

Prediction 45

"Incessant corrosive

Space debris

punches Mars"

The Chronicles of Iris

...futuristic predictions

Chronicle (11152018-T1)

November 15, 2018

Prediction 46

"Unyielding Space debris bashes Neptune"

The Chronicles of Iris

...futuristic predictions

Chronicle (11152018-U1)

November 15, 2018

Prediction 47

"Mammoth space debris plummets Jupiter"

The Chronicles of Iris

...futuristic predictions

Chronicle (11162018-V1)

November 16, 2018

Prediction 48

"Annihilating space weather permeates the Milky Way"

The Chronicles of Iris

...futuristic predictions

Chronicle (11162018-W1)

November 16, 2018

Prediction 49

"Spacecrafts operating on Quantum energy"

The Chronicles of Iris

...futuristic predictions

Chronicle (11162018-X1)

November 16, 2018

Prediction 50

"Spacecrafts designed for Quantum Physics"

Purchase my other chronicled

prediction books from Amazon

books & Kindle

Watch & subscribe to
The Chronicles of Iris Podcast

The Chronicles of Iris..
YouTube

www.thechroniclesofiris.com

www.thechroniclesofiris.com

www.ingramcontent.com/pod-product-compliance
Lightning Source LLC
Chambersburg PA
CBHW071123210326
41519CB00020B/6397